OSHKOSH TRUCKS

75 YEARS OF SPECIALTY TRUCK PRODUCTION

David Wright
Clarence Jungwirth

Motorbooks International
Publishers & Wholesalers

First published in 1992 by Motorbooks International Publishers & Wholesalers, PO Box 2, 729 Prospect Avenue, Osceola, WI 54020 USA

© David K. Wright, 1992

All rights reserved. With the exception of quoting brief passages for the purposes of review no part of this publication may be reproduced without prior written permission from the Publisher

Motorbooks International is a certified trademark, registered with the United States Patent Office

The information in this book is true and complete to the best of our knowledge. All recommendations are made without any guarantee on the part of the author or Publisher, who also disclaim any liability incurred in connection with the use of this data or specific details

We recognize that some words, model names and designations, for example, mentioned herein are the property of the trademark holder. We use them for identification purposes only. This is not an official publication

Motorbooks International books are also available at discounts in bulk quantity for industrial or sales-promotional use. For details write to Special Sales Manager at the Publisher's address

Library of Congress Cataloging-in-Publication Data
Wright, David.
    Oshkosh Trucks / David Wright, Clarence Jungwirth.
      p.   cm.
    Includes index.
    ISBN 0-87938-661-4
    1. Oshkosh Truck Corporation— History.   2. Truck industry—United States—History.   I. Jungwirth, Clarence. II. Title.
HD9710.35.U64O839   1992
338.7'629224'0973—dc20   92-28520

Printed and bound in Hong Kong

The front and back cover color illustrations were specially commissioned by the Oshkosh Truck Corporation and are used with their permission.

On the back cover: Fifty years separate a Model A carrying hemp for rug making and an LVS (Logistics Vehicle System) with sand tires.

On the title page: A fleet of W-series mine trucks is delivered by rail to ALCOA.

Contents

Introduction	**From Four-Wheel-Drive to Ten-Wheel-Drive**	7
Chapter 1	**Four-Wheel-Drive Pioneers 1915-1938**	9
Chapter 2	**The Growth Years 1939-1986**	19
Chapter 3	**All-Wheel-Drive Leaders 1968-1992**	29
Chapter 4	**Photo Album**	45
Afterword	**Looking to the Future**	127
	Index	128

Part of the company's first factory.

Introduction

From Four-Wheel-Drive to Ten-Wheel-Drive

In the early 1900s, shade-tree mechanics all over the United States tinkered in barns and sheds with the idea that a vehicle could run on its own power. They dreamed of a vehicle that would go where the usual forms of mechanized transport—trains and boats—could not. It would run wherever it was pointed.

For several years, a vehicle capable of going almost anywhere remained a dream because early roads were a nightmare. Created from dirt and gravel with no thought given to drainage, early 20th century roads in the United States evolved from animal trails, from the paths of Native Americans, and from pioneer wagon tracks. In warm weather they were dusty and at other times they were snow-covered or cratered with mud, slush, and standing water. American productivity declined each year with cold or wet weather and spiraled upward again in late spring.

This problem was attacked from two angles. The first concrete street in North America was laid down in Bellefontaine, Ohio, in 1891. Concrete and asphalt eventually found favor and spread. But it would take decades for pavement to cover most streets and highways and, after several years of use and weather, repaving was necessary. Clearly, there should be another answer.

A solid alternative surfaced in Clintonville, a small farming and manufacturing town in central Wisconsin. The Wisconsin Duplex Auto Company was organized to develop and produce a four-wheel-drive vehicle. Like paved roads, the vehicle was a gradual but resounding success, improving productivity and helping tame what was still a raw and rugged country.

W. R. Besserdich and B. A. Mosling, co-founders of what is now Oshkosh Truck Corporation, looked at transportation problems in different ways. Besserdich, the mechanic, believed that power to all four wheels was the answer to the problem of traveling over the awful roads. Mosling, the merchant, realized that once roads were developed, a new, nationwide era in transportation and productivity would evolve.

Making productive transportation equipment that goes on and off road has been the Oshkosh Truck Corporation hallmark for all seventy-five of its years. From the first prototype (a four-wheel-drive truck named *Old Betsy* that is now on stately display at the company) to ten-wheel-drive military vehicles that provide the US Army with superior mobility and efficiency, the company has filled a unique niche in the story of American transportation.

Today, Oshkosh Truck Corporation manufactures a wide variety of specialized trucks and transport equipment. Products are engineered for specific market niches where a unique, innovative design will outperform general-purpose equipment. The company's major products include heavy-duty commercial and military trucks, trailers, proprietary drive components, and motorized chassis for the motorhome, bus, and walk-in delivery van markets.

Through the years, Oshkosh Truck has pioneered high-performance innovations that resulted in new and better ways of transporting goods under adverse conditions throughout the world. William R. Besserdich and Bernhard A. Mosling would be proud of the accomplishments of Oshkosh Truck and the creativity and dedication of its people as it completes seventy-five quality-focused years.

Old Betsy, *the first Oshkosh truck.*

Chapter 1

Four-Wheel-Drive Pioneers 1915-1938

From *Old Betsy* to the J-Series Trucks

Had the American automobile industry shown a little more imagination, there might be no Oshkosh Truck Corporation today.

William R. Besserdich and Bernhard A. Mosling, co-owners of two patents improving four-wheel-drive capability, approached Case, Ford, Jeffery, Kissel, Packard, Studebaker, and others in 1916 about producing a four-wheel-drive car or truck using their component designs. Neither man at the time saw himself involved in vehicle manufacturing. One by one, the car companies rejected their queries, so Besserdich and Mosling decided that the only way to get their innovative ideas into production was to build their own four-wheel-drive vehicle.

Four-Wheel-Drive Pioneers

The two believed they were onto something, despite the rebuff by the automakers. One patent covered the transfer of power equally to all four wheels by means of an automatic locking differential in the transfer case. The transfer case is a component between the frame rails behind the transmission that transfers power to the front and rear axles. The other patent greatly improved the steering and drive capability of the front driving axle.

Besserdich worked on designing a prototype vehicle while Mosling hit the road to sell stock in the Wisconsin Duplex Auto Company. The firm was incorporated on May 1, 1917. That happened to be the day before the federal Conscription Act was passed, allowing the government to draft young men for war in France. Back in Clintonville, Wisconsin, work began with $250,000 of capital realized from stock sales.

"W.A. Besserdich," stated a company flyer, "is so well known throughout the state and elsewhere . . . that little need be said of him." B. A. Mosling, several years younger than his partner, "has been in the mercantile business for more than 20 years and has never known a failure." Mosling had been involved in the lumber and retailing businesses. A longtime Oshkosh Truck Corporation employee, now retired, recalls that Mosling's energy seemed boundless. While others sat through meetings, Mosling

William R. Besserdich.

Bernhard A. Mosling.

The automatic locking differential.

The front driving axle.

The Oshkosh Motor Truck Manufacturing Company.

The Oshkosh Model A.

would pace the floor ceaselessly, overflowing with thoughts and ideas. Routinely, his days began at 7:30 am and ended around 11 pm.

Besserdich was listed as president, Mosling as secretary and manager. Attorney A. S. Larson was vice president and J. P. Mosling, father of B. A. and a successful retailer and lumber manufacturer, signed on as treasurer. The company offered 5,000 shares of stock at $100 per share.

Old Betsy

As money was raised, construction began on the prototype vehicle. *Old Betsy* was crafted by a Milwaukee machine shop and powered by a four-cylinder LeRoi engine. The gasoline engine was connected to a three-speed transmission and the 1 ton capacity truck weighed in at 3,280lb. This initial truck ran on 32x4in Firestone gum-dipped tube tires; like all Oshkosh trucks, pneumatic tires were standard equipment. *Old Betsy* incorporated the patented, automatic positive-locking center differential in the transfer case. The prototype truck also featured the carefully crafted, patented front-axle steering pivots. The truck's major accomplishment was showing the advantages of these four-wheel-drive components, which helped sell more shares in the company.

The truck did its job. At least half of the stock sold quickly, much of it to financial interests in Oshkosh, forty-seven miles south of Clintonville. The young corporation moved there late in 1917 and became the Oshkosh Motor Truck Manufacturing Company. A building was leased from a supplier on a site which is now the campus of the University of Wisconsin-Oshkosh and production began.

Model A Trucks

The first production truck was the Oshkosh Model A, which featured a door on each side of the cab. (Most trucks with cabs in those days required entry from the passenger's side. The steering wheel blocked entry from the driver's side.) The windshield and windows were adjustable for ventilation. The four-wheel-drive truck produced 72hp, thanks to the fact that the Herschel-Spillman four-cylinder engine heated the fuel at three different points to get the most from the low-octane gasoline of the time. The Brown-Lipe Model 35 transmission featured four forward speeds and reverse. The frame was fabricated by A. O. Smith Company.

Standard features included 36x6in Goodyear diamond-pattern pneumatic cord tires on demountable rims, Allis-Chalmers electric starting, two headlights and a taillamp, a speedometer, and an electric horn. The Model A offered thermo-syphon cooling, set-back axles with artillery-style spoke wheels, and a five-spoke steering wheel. The standard color was brown and the approximate price was $3,500. The Model A was one of more than 100 different makes of vehicles manufactured in Wisconsin at the time, according to 1920 US Census abstract figures. What made the Oshkosh truck special was four-wheel-drive.

It's easy to forget, seated in today's human-engineered cabs, just how tough driving a truck used to be. A wooden bench covered in little except a mat of horsehair and a thin sheet of leather was a virtual springboard. It wasn't unheard of for drivers to suffer cracked ribs as they were pitched by rutted roads around the cab interior if they weren't careful. Drivers also could break thumbs as the steering wheel whirled crazily from the kickback of the live front axle dropping in a rut.

Early Oshkosh trucks were fully suspended and seldom got stuck. Under rugged conditions, drivers reported being able to average an amazing "14 to 20 miles an hour" traveling between Oshkosh and Milwaukee because of the truck's all-wheel-drive capability.

Oshkosh solicited reports from owners and received hearty endorsements. J. H. Schneider, owner of a grain mill, wrote, "We have driven our truck since March 21st and it has run 2,500 miles, this being through the worst kind of roads. Our trips to Reedfield and other points . . . are over very bad roads, especially during the Spring of the year. Our Oshkosh 4-Wheel-Drive negotiated these roads with over-capacity loads, without any trouble, passing many trucks on the road that were stalled in the mud for many hours."

Enough people agreed with Mr. Schneider to encourage the company to look for larger quarters. While Americans in 1920 were listening to the very first commercial-radio broadcast or reading Sinclair Lewis's bestselling *Main Street,* The Oshkosh Citizens Building Company made possible the construction of a structure south of the city of Oshkosh. That building, with many improvements and

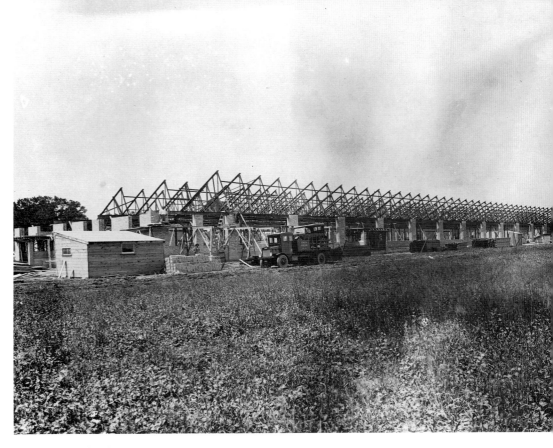

The south-side plant under construction.

The home of Oshkosh Truck.

The Model F.

expansions, has been occupied continuously by Oshkosh Truck since December 1, 1920. About two dozen employees began to produce the Model A in the new facility on January 1, 1921.

The Model A could carry 2 tons, more than most conventional trucks, and with good reason. B. A. Mosling saw very early that large-capacity models would reduce labor and transportation costs. The Model B, offered soon afterward, was rated at 3.5 tons capacity and was quickly followed by the 5 ton Model F.

Sales climbed from seven units in 1918 to fifty-four in 1919 to 142 in 1920. However, due to a postwar depression, profits lagged. The market was also complicated by the fact that the US government gave away surplus trucks to municipalities after World War I. B. A. Mosling succeeded Besserdich as president in April 1922. (While Besserdich was instrumental in getting the company started with his engineering talents, he never played an active role in the operation of the company once production began.) Sales in 1921 totaled just sixty-two trucks, followed by nineteen, sixteen, and twenty-four in the three tough years that followed.

The city of Oshkosh at the time was occupied by many first- and second-generation German-Americans, as well as immigrants from many other European countries. These people brought craftsmanship and a willingness to work—as well as sincere loyalty—to the small truck-building company on the south side of town. The employees worked long hours to support their large families. Occasionally, they slept overnight in the factory when the spring thaw caused Lake Winnebago and the Fox River to flood some of the roads of the low-lying community.

At about the same time, 1921, *The Oshkosh Daily Northwestern* newspaper reported that, as part of an Oshkosh Motor Truck advertising film, a Model A truck was repeatedly driven up the steps of the local high school. "Miss Blanche Rahr of this city at the wheel piloted the big machine . . . in a fashion that won the applause of the spectators who had gathered to witness the stunts." The same feat was performed later on the steps of the local library. The company advertised that an Oshkosh truck "Goes Anywhere The Wheels Can Touch the Ground," a slogan with which local eyewitnesses would agree.

Model H Trucks

The Model H, introduced with double-reduction axles in 1925, sold briefly with a four-cylinder engine but was most popular with six cylinders. The resulting power and traction were superior to earlier four-wheel-drive trucks and far ahead of anything with conventional two-wheel-drive. The new truck found favor with municipalities in the areas of snow removal and road construction, two tasks that grew throughout the 1920s. Early employees recall groaning at the sight of snowflakes fluttering past plant windows, but the flakes cheered B. A. Mosling, who called them "pennies from heaven." Heavy snows meant increased truck sales.

Snow removal required special techniques. Since hydraulic systems were several years away, one person drove while the other manually cranked a big V-plow and a passenger-side wing plow over or around fire plugs, street signs, and mail boxes. Like Oshkosh trucks before and since, the Model H left the factory with pneumatic tires and was completely painted and ready for work. It gained the company loyal buyers for more than five model years.

The economy remained so depressed that, by the mid-1920s, a number of larger and more well-known truck and automobile manufacturers had fallen by the wayside. For the company, the Model H returned just enough revenue to keep it afloat. In 1930, however, with the country in the middle of the Great Depression, the company was reorganized as Oshkosh Motor Truck Company, Incorporated, with R. W. Mackie installed as president. B. A. Mosling, a salesman at heart, focused on the challenge of selling trucks during a period when individuals, corporations, and municipalities had few dollars.

Models FC and FB Trucks

The Depression bottomed out in 1932, the same year Oshkosh Truck introduced several new models. These were the FC and the FB, with capacities of as much as 44,000lb gross vehicle weight (GVW). They were powered by six-cylinder Hercules gasoline engines ranging from 102hp to 200hp. Transmissions could be had with from four to twelve speeds, all of them driving through the popular double-reduction axles. The first Oshkosh Truck with a diesel engine was offered with a Cummins in 1935.

As road surfaces improved and winter traffic increased, Oshkosh trucks were in demand for snow-removal uses. Municipal snow-removal service marked a major change as before the late 1920s, winter roads rarely were plowed. Many cars spent winters on blocks, since the only radiator coolant besides water was alcohol, which boiled away quickly once the engine warmed up.

Not only were trucks faster than the crawler tractors commonly used for snow

The Model A. Miss Blanche Rahr driving up the steps.

The Model H.

The Model FB.

removal in rural areas, they also were less destructive to road surfaces. And because they were faster, they could hurl snow farther and get the job done quicker. With hydraulics now controlling plow and wing movements, and with more miles of road to clear with each succeeding winter, four-wheel-drive trucks gained favor over crawlers in highway departments throughout the United States. In better weather, the dump-bodied trucks were used to haul gravel for road repair and construction.

The Model TR.

Model TR Trucks

The innovative Model TR, introduced in 1933, was a diversification by the company into a non-snow removal, non-municipal market. This vehicle, the first rubber-tired earthmover ever built, was a large, four-wheel-drive vehicle for use with bottom-dump trailers, self-loading scrapers, or dozer blades. Goodyear made special tire molds, which were larger than anything available at the time.

The TR had four-wheel steering. When the steering wheel was turned clockwise, the front wheels turned to the right and the back wheels turned to the left. This allowed the big off-road machines with their four-wheel steer to turn 180 degrees in a diameter measuring just 31.5ft.

TRs were advertised as utility vehicles and proved popular with contractors working on dam, canal, and airport construction all across the country. These dramatically new vehicles, capable of speeds up to 35mph, were among the first attempts by the company to diversify the product line. Their setback front axles gave them an unusually modern look and enabled them to achieve approximately equal loading on front and rear axles. The TR was evaluated by the US Army at Ft. Bragg, North Carolina, for pulling a 155mm gun. Although it ran rings around contemporary crawlers and tractors, the military declined to buy TRs because of their high profiles and the vulnerability of rubber tires. The vehicles found favor in England, where more than 100 were sold to mining operations.

J-Series Trucks

A contrast to the TR was the J-Series. These trucks were introduced in 1935 and were another attempt to broaden the Oshkosh line. The vehicles had classic 1930s automotive styling: one-piece windshields, new cabs, and gently rounded fenders and grilles. Capacities of the four-wheel-drive Js were between 2 tons and 3½ tons.

The J-Series.

During this time there were also Oshkosh Express models which were an attempt to compete in the rear-drive market against mass producers. The company soon dropped these models to concentrate on the more specialized all-wheel-drive market.

Leading Oshkosh truck markets in the 1930s, besides Wisconsin, included New York, the New England states, New Jersey, and the state of Washington. Wisconsin and New York had similar needs. Milk trucks required freshly plowed roads to reach thousands of remote dairy farms every day, all winter long. Farmers had enough influence to make sure state and local officials budgeted every year for the best in snow-removal equipment. The same year a national minimum wage law was first enacted, 1938, Oshkosh trucks were earning their keep by clearing paths for the next day's milk in major dairy states. Truck prices that year ranged from $2,885 to $13,500.

The W-Series.

Chapter 2

The Growth Years 1939-1986

From the W-Series Trucks to Defense Contracts

Despite the economic hardships of the 1930s, Oshkosh Truck had seen the wisdom of expanding its engineering department and greatly improving its components. That activity paid off in 1939 when the W-Series vehicles were introduced. These trucks, with gross vehicle weights of from 18,000 to 44,000lb, did the same sorts of tasks as their predecessors, but did them better. Attention had been paid to styling, driver comfort (hydraulic power steering was a popular option), increased power, more capacity, and availability of either gasoline or diesel power across the product line. When World War II broke out, the company had already fully developed the W-Series trucks, which were very suitable for the military as snowplow and dump body vehicles.

W-Series Trucks

Early W-Series trucks were easy to identify. They were adorned with an Indian-head grille and offered a more modern cab with the V-type windshield. The first W-Series to see military duty during the conflict was the Model W-700, chosen by the US Army Corps of Engineers. The engineers employed rotary snow-blower equipment to keep Army Air Corps runways free of snow around the world, along with a number of the trucks configured as wreckers. The rotaries were powered by a 175hp Climax engine mounted on the back of the truck. Both the Climax and the six-cylinder, 112hp Hercules RXC truck engine powering the truck were gasoline-fueled.

The Model W-700.

A substantial number of snow removal trucks also were used to keep United States roads clear so that wartime production schedules at domestic factories could be met. Because Oshkosh Truck had been prudently managed, it was able to finance its war-production program without borrowing. And because the War Department was using virtually the same products the company would have produced in peacetime, the company was able to steadily improve the design and quality of all of its products.

The company received many awards from the government for excellence in wartime production. B. A. Mosling in 1944 resumed the title and responsibilities of president. He remained at the helm until 1956, when he was succeeded by his son, John.

As the war was ending, the Model W-1600 made the transition from drawing board to reality. These vehicles were 6x6s,

The Model W-1600.

B. A. Mosling.

with all three axles being driven. They were designed initially for use off-road in America's oil fields and as heavy-duty tractors pulling trailers loaded with machinery and other heavy cargo.

This line was complemented in 1947 by the Model W-2200, produced through 1955. Larger, faster, heavier, and more powerful than their competitors, the W-2200s were 4x4 vehicles that ran Buda and Hall-Scott gasoline or Cummins diesel engines. The larger trucks could be equipped with larger plows and wings than virtually any other truck available at the time. And with a modern hydraulic system to operate the attachments, one person could clear more snow faster than any two people in any other type of vehicle (although one person was so busy that two people was the norm). The versatility of Model W-2200 tractors pulling trailers was proven after they were purchased in quantity by major sugar companies to haul the harvest from plantations to processors, and by many mining companies to haul ore.

Ready-Mix Market Thrust

Meanwhile, other exciting events were occurring. The millions of men who mustered out of World War II got jobs, married, and started families, which meant they needed housing. Curbs, gutters, slabs, basements, driveways, and streets were required at a much greater rate. In response to the building frenzy, Oshkosh produced a revolutionary concrete carrier called the Model 50-50 in 1955.

This 4x4 truck was an important innovation in the ready-mix concrete industry because of its increased capacity and because all-wheel-drive provided better service to contractors: getting into jobs and out of job sites without getting stuck. Oshkosh trucks became favorites in Florida, Indiana, Michigan, and Ohio—the nation's leading ready-mix markets.

The 50-50 was the first truck ever created specifically for the hauling of concrete. It had a set-back, driving front axle that enabled the vehicle to carry 50 percent of the gross weight on the front wheels and 50 percent on the back, hence the model name. Concrete producers were able to haul more legal payload and have more off-road mobility with these specialized trucks than they could with conventional over-the-road vehicles.

The gasoline-powered 50-50 was so successful that a diesel-powered model, the Model 45-55, was brought out the same year. Whereas the 50-50 had a rear axle with 18,000lb capacity, the 45-55 featured a rear axle rated at 23,000lb capacity. Like all Oshkosh trucks, the driving front axle and transfer case used on the vehicle were designed and manufactured by the company. Gross vehicle weights for the two trucks were 36,000lb and 41,000lb, respectively.

The Model 50-50.

The Model W-2200.

The Model 1832.

The F-Series.

The first tandem-axle, creating a 6x6 truck, came out in 1956 and was designated the Model 1832. Again, these were based upon the 50-50 design, the model designation standing for weight ratings of 18,000lb on the front axle and 32,000lb on the tandem. During the early 1960s, axle-forward models were introduced that were different from "long-nose" set-back axle models and became the F-Series. Axles were moved slightly forward for maximum payload in states with bridge formulas for weight distribution requirements. The set-back-axle cantilever models became the C-Series.

Construction surged in the 1960s, creating a demand for larger concrete carriers. A significant design feature of these trucks was the introduction of a large driving front axle that could carry 21,000lb and later 23,000lb. The cantilever-engine design C-Series was replaced by the F due to a change in mixer design that altered the weight distribution, making the axle-under-the-cab location less beneficial.

New federal brake standards also became a factor in favoring the Oshkosh axle-

The D-Series.

forward position in order to accept more weight during severe braking. Available in 6x6, 8x6, 10x6, and 10x8 drives, F-Series concrete carriers were seen providing superior customer service with their all-wheel-drive capability. They were also used extensively for concrete-block hauling and for utility construction work. High-capacity 16 cubic-yard trucks became available with tandem driving front axles. They were designated the D-Series and were especially popular in Michigan.

The C-, F-, and D-Series trucks set the standard for concrete hauling, but the factory did not rest on its laurels. Aware that drivers of concrete carriers had to back into congested job sites to discharge concrete from the chute, Oshkosh Truck developed a forward-placement concrete carrier.

Introduced as the B-Series in 1975, the concrete carrier had a one-person cab placed over the front axle with the engine mounted in the rear. With all mixer controls in the cab, the operator could drive directly to the spot where the contractor wanted the concrete placed and discharge the concrete as the chute was controlled by

The B-Series.

The S-Series.

the driver from inside the cab. This eliminated the high cost of having to wheelbarrow the concrete. Equally important was the safety factor. Forward placement eliminated the possibility of backing into a basement excavation. The vehicle also saved money and time by placing the concrete while moving along a foundation or stretch of curb.

The next Oshkosh advancement was the availability of a chassis and mixer as a complete package. This series, designated S, first came out in 1982 and featured an Oshkosh brand mixer. The S-Series gave contractors the convenience of purchasing and receiving field support from a single source. It also eliminated delays caused by moving a finished truck elsewhere for mixer installation.

Expansion of the Truck Line

The ready-mix business expanded year after year for Oshkosh Truck, while new markets called for innovative designs.

A large and unusual descendant of the F-Series was the J-Series (which bore no relationship to the J-Series vehicles of the 1930s). In 1974, the company debuted a pair of J-Series trucks for oil-field work, the Desert Prince and the Desert Knight. These huge, six-wheel-drive trucks ran with 325hp to 485hp diesel engines, 2,000 sq.-in. radiators for desert cooling, and large balloon tires for flotation over sand. On some models, the tubular front bumpers held fresh drinking water! Most were used in the deserts of the Middle East and China, but some saw service in other parts of the world as tractors in heavy-haul applications on highways.

Oshkosh Truck also introduced the heavy-duty R-Series. These conventional 6x4 trucks and tractors were more ruggedly designed than domestic trucks so

The J-Series.

they could withstand severe operating conditions found in Australia, Africa, the Middle East, and elsewhere, where road conditions were less than ideal.

Also manufactured for the worldwide over-the-road truck market, the company designed a handsome cab-over-engine model in the 1970s designated the E-Series. Like the R-Series trucks, the E-Series was Caterpillar powered. It was assembled in kit form in Australia and was manufactured in South Africa for a time into the early 1980s.

Defense Leads Growth

While it met civilian needs with a succession of innovative new models, the company kept an attentive eye on military and other governmental requirements. As early as 1960, the military was a major customer of the company and would continue to be a significant factor in the company's growth and technological advancements in the years to follow.

This first major defense contract since World War II came about as a result of the Cold War. The United States knew that the Soviet Union was capable of launching a surprise air attack on North America. To prevent such an attack, the United States and Canada strung a line of distant early warning (DEW) radar stations across Canada and Alaska. This web of radar would alert the military, especially the US Air Force.

The Air Force had several Strategic Air Command (SAC) bases in the northern tier of states with B-36 and, later, B-52 bombers poised to retaliate. But since most bases were deep in the snow belt, the military needed a method to open runways immediately and to keep them open, no matter how much snow should fall. Oshkosh Truck created a revolutionary concept in a new model, the WT-2206. These large, heavy-duty trucks with 325hp Hall-Scott engines and Allison TG 602-RM automatic transmissions were capable of operating at 55mph while plowing in formation, pushing snow in a wide, one-way arc far past runway lights. The high-speed truck was half the equation. The other half involved a plow that was as innovative as it was simple.

Before Oshkosh Truck addressed the challenge, snow removal vehicles moved down runways, then lifted their conventional blades and returned to the starting point so that all of the snow could be pushed in the same direction. But Oshkosh specified a big rollover plow that could be raised and rolled over. The trucks could then simply turn around at the end of the runway and make another pass, since all of the snow was now being pushed in the

The E-Series.

direction of the first pass. Oshkosh won the contract to produce more than 1,000 of the vehicles, which could also be equipped with rotary snowblowers.

The WT-2206 also showed commercial airport management the benefits of high-speed snow removal. Sales grew as airport managers realized they could remain open during most storms, reducing disruption to airline schedules.

The high-performance Oshkosh Air Force trucks seldom needed more than routine maintenance. But as time passed and new technology became available, the Air Force worked with Oshkosh to remanufacture the older vehicles, upgrading them to current state of the art. For example, through time all the units were converted to diesel power from gasoline. Trucks and plow equipment were remanufactured for less than 60 percent of the cost of a new vehicle and included a new-truck warranty.

Oshkosh trucks, some as far back as the 1940s, were upgraded through remanufacturing three or four times over a thirty-year period, reappearing each time as more modern trucks. By the fall of 1986, this program had saved taxpayers an estimated $83 million.

The R-Series.

The Model WT-2206.

The US Navy MB-5 Aircraft Rescue and Fire Fighting (ARFF) truck.

Chapter 3

All-Wheel-Drive Leaders 1968-1992

Trucks for All Uses

The US Navy MB-5, with its aluminum body, was an aircraft rescue and firefighting (ARFF) truck built under contract in 1968. Actually, the first ARFF vehicle built by Oshkosh was a W-Series truck delivered to the US Coast Guard in 1953, but the MB-5 launched Oshkosh into a worldwide leadership role that continues today. It carried 400 gallons of water which, when mixed with a foam concentrate, expanded to 5,000 gallons of extinguishing foam and discharged from a roof turret. The vehicle also is significant because a dozen of the units were used at sea, putting out many fires on the flight decks of aircraft carriers.

The original MB-5 order was followed by another for 300 and led to an order for seventy-three MB-1s of 1,000 gallon capacity for the Navy in 1971. The MB-1 also was available in civilian versions, designated the M-Series.

Other crash-truck configurations followed. The P-4, a 6x6, could carry 1,500 gallons, and 542 were sold to the US Air Force in the early 1970s. P-4As were sold to the US Navy and to the Australian Air Force. Much larger was the 66 ton P-15, introduced in 1977. This massive firefighting vehicle was powered by a *pair* of 492hp Detroit Diesel V-8s, which furnished thundering acceleration to all eight wheels. It carried 6,000 gallons of water which, when mixed with concentrate, made 60,000 gallons of fire-suppressing foam. A later US Air Force contract was for the P-19 aircraft rescue and firefighting vehicle, which was

The US Navy MB-1.

The US Air Force P-4.

The US Air Force P-15.

The US Air Force P-19.

awarded in 1984. Some 715 of the 1,000 gallon capacity P-19s were delivered.

Expertise in years of manufacturing various aircraft and rescue firefighting vehicles resulted in the introduction in the late 1980s of an 8x8 DA articulated airport rescue and firefighting vehicle. It was derived from an articulated, high-mobility tactical cargo truck built for the US Marine Corps and was produced in 1,500 and 1,800 gallon versions. With minor modification, these were sold to commercial airports.

Aircraft tow tractors were another type of vehicle manufactured by the company for the US Air Force. In 1968, Oshkosh Truck designed and built the U-30, which was used to tow the giant C5A cargo aircraft. A total of forty-five were built. The MB-2 was a smaller tow tractor built for the US Air Force from 1968 to 1977. A total of seventy-two of these units were produced.

The company in 1976 won a major US Army contract that led to delivery of 744 tractors used to pull trailers loaded with tanks or heavy equipment. This was the company's first US Army contract. Designated the M-911 Heavy Equipment Transporter (HET) by the Army, the vehicle was unique in that the bidding for the contract required it to be based on a commercial tractor which had to be currently in production. The Oshkosh HET was derived from an F-Series truck. The M-911 is still produced today to meet export requirements.

The 8x8 DA Aircraft Rescue and Fire Fighting vehicle.

The US Air Force U-30.

The US Army M-911 Heavy Equipment Transporter (HET).

The US Air Force MB-2.

The US Air Force 40K.

In 1979, Oshkosh Truck Corporation received a contract to produce aircraft loaders for the US Air Force. More than 100 aircraft loaders called the 40K were delivered. The 40K designation indicated that the vehicle had a 40,000lb lifting capacity. It had four axles, only one of which was driving.

Heavy Expanded Mobility Tactical Trucks

In 1981, the company won its largest government contract to date. Oshkosh Truck was the successful bidder to construct Heavy Expanded Mobility Tactical Trucks (HEMTT), the trucks that proved crucial for ground support during Operation Desert Storm in 1991. General H. Norman Schwarzkopf, Commander in Chief—US Central Command during Operation Desert Shield/Desert Storm—told the House Armed Services Committee that, without trucks, "we never would have had the supplies far enough forward to go ahead and launch the war . . . I am a great believer in the HEMTT. . . ."

Secretary of the Army Michael P. W. Stone personally visited Oshkosh Truck in 1991 to thank employees for their support during the Gulf War, where the HEMTT played such an important role, especially in refueling other vehicles. "If we're going to retain quality soldiers, we need to obtain quality equipment," he said. "And for the past 10 years, you've been delivering one of the finest pieces of equipment that we have in the United States Army." Another visitor, General William G. T. Tuttle, Jr., Commander of the US Army Materiel Command, told employees, "The Army, Air Force, Navy and Marine Corps did a hell of a job over there, but that job couldn't have been done without you."

More than 13,000 HEMTTs have been delivered as of this writing. There are five models of the 8x8 truck, including two cargo trucks, a tanker, a tractor, and a recovery vehicle. The tractor version pulls the Patriot missile launcher. Power comes from a 445hp Detroit Diesel V-8 engine via a four-speed Allison HT-740 automatic transmission.

Ongoing Military and Civilian Service

As mentioned earlier, the company also has served the US Marines. The corps took delivery of 1,400 Logistics Vehicle Systems (LVS) trucks beginning in 1985, which feature center articulation for increased mobility over soft and uneven terrain. The vehicles have several different rear sections that can be uncoupled and interchanged. Uncoupling also permits lifting by helicopter.

A recent refueler contract was for the R-11, a 6,000 gallon capacity truck introduced in 1988 for the US Air Force. Powered by a 250hp diesel engine with a five-speed automatic transmission, the vehicle is designed to fuel or defuel aircraft. The R-11 has a non-driving front axle and a driving tandem in the rear. A total of 1,267 were built. The R-11, like other defense vehicle products, found use with many nations worldwide.

Snow removal continues to be an important market for Oshkosh, and the company is currently, in the 1990s, providing a new-generation vehicle for the US Air Force. More than 300 of these multi-purpose units will be produced for the military. They are unique because they can be used either as a snowblower or blade plow truck. The blower head and blade plow are interchangeable. The 4x4, four-wheel steer units can blow 3,000 tons of snow per hour. The vehicles are also sold to commercial airports where many also are equipped with rotating brooms to sweep light snow and dirt from runways.

A contract for the new US Army truck known as the Heavy Equipment Transpor-

The US Army Heavy Expanded Mobility Tactical Truck (HEMTT).

US Army HEMTT model.

The US Marines Logistics Vehicle System (LVS).

The US Air Force R-11.

The US Air Force snow removal unit.

ter (HET) M-1070 was awarded in January 1990. A total of 1,044 vehicles will be delivered. This tractor is different from the M-911 HET in that it has additional hauling capacity, 500hp, a five-person cab, 8x8 drive, central tire inflation, and a steerable rear axle. The vehicle's main mission is the hauling of the 70 ton M1A1 main battle tank.

Another major US Army contract was awarded to Oshkosh Truck in September 1990, this one for 2,626 of the Army's Palletized Load System (PLS) vehicles, which will revolutionize the service's logistics. These vehicles employ a hydraulic load-handling system with a hook that loads a flatrack cargo bed onto its back in a single motion. The truck has a carrying capacity of 16½ tons and also pulls a trailer with a 16½ ton payload flatrack mounted on it. The highly mobile, cost-reducing, and time-saving vehicle is a 10x10 with a 500hp diesel V-8, automatic transmission, central tire inflation, and rear-steer axle, in addition to the tandem-steering front. The Army estimates lifecycle cost savings of approximately $600 million from its fleet of PLS vehicles.

Like the HET, the central tire inflation system on the PLS allows the operator to inflate the tires for highway use and deflate them for travel off road in soft ground. This can be accomplished while the truck is

The US Army Heavy Equipment Transporter (HET).

The US Army Palletized Load System (PLS) vehicle.

Concrete carrier.

Snow-removal equipment.

moving. Also like the HET, the last axle steers for a decreased turning radius and improved maneuverability.

Oshkosh Truck in the Future

As Oshkosh Truck continues to pursue military equipment opportunities in the future, it will also be aggressively investing its specialized vehicle design and manufacturing resources in commercial market areas. Advanced generations of snow-removal equipment, concrete carriers, aircraft rescue and firefighting units, and other specialized vehicles are continuously being developed and introduced to better serve customers.

An example of a new commercial product is a unique combination recycler/refuse collection truck that recently went into production. It's the type of innovative, cost-saving vehicle that can be expected from Oshkosh in the future. Because the

Concrete carrier.

The recycler/refuse collector.

Aircraft rescue and fire fighting units.

A specialized chassis.

A military trailer.

truck collects both refuse and recyclables on the same run, it can replace two single-function trucks, saving time and money.

The stripped chassis business is a significant, relatively new contributor to Oshkosh sales. Chassis are built for motorhomes, walk-in delivery vans, and buses.

In 1990, the company further diversified with the acquisition of a trailer manufacturer in Florida and is now a builder of a broad line of trailers, including many specialized models such as fruit haulers and a variety of military trailers.

A fruit-hauling trailer.

Chapter 4

Photo Album
Seventy-Five Years of Oshkosh Trucks

Company co-founder B. A. Mosling stands in front of one of the first Oshkosh trucks, demonstrating its four-wheel-drive climbing ability.

The prototype Model A with a number of original Oshkosh Truck employees posing on the truck body. The company was named the Wisconsin Duplex Auto Company at the time.

Pictured are the manufacturing areas in the company's first factory. The leased building was on the site which is now the University of Wisconsin-Oshkosh.

This truck is equipped with a mechanical hoist with power take-off driven from the side of the transmission. Many early trucks did not have cabs because most construction work was performed in fair weather.

These parts make up the patented steering mechanism of early Oshkosh trucks.

A Model B destined for flour delivery from the mill to the grocery store. The city of Oshkosh in 1920 had 40,000 residents and approximately 100 grocery stores.

Cab interior of a Model H truck.

Next page
Top
This "Interurban Express" has chains on the front wheels for extra traction. Many smaller communities surrounding larger cities in Wisconsin were serviced by the motor-driven "wagon trains."

Center
A Model B tows a portable concrete mixer powered by a steam engine. The loading hopper is in the bed of the truck. Concrete was beginning to be used in place of wood in sidewalk construction.

Bottom
A fleet of fifteen Model Bs, ready for delivery to the Standard Oil Company.

An early "Oshkosh Express" model delivers coal in the 1920s. "Express" models had non-driving front axles.

The Wisconsin Hemp Company of Brandon, Wisconsin, used this Model A to haul marsh grass to its manufacturing plant. Marsh grass was used to make low-cost rugs and other products.

Previous page
Top
This mobile grocery store provided all-wheel-drive service to rural shoppers.

Bottom
This Oshkosh Model H used by the Waupaca (Wisconsin) County Highway Department for plowing country roads was capable of speeds of 35mph with its six-cylinder gasoline engine and seven-speed transmission.

This Model B fire truck is shown in front of its fire house. Notice the plush upholstery, a rarity for trucks at the time.

This Model H has dual rear tires. Duals on the rear allowed the truck to handle heavier loads. Once this trend for larger payloads began, heavier axles and wheels resulted in higher gross-weight ratings.

Previous page
Top
A Model H hauls logs on a pole trailer. Four-wheel-drive allowed loggers to get their product out of the woods.

Bottom
The Model H fire truck and trailer with tiller steer has solid tires. Fire departments may have insisted on solids in the 1920s to eliminate any chance of a flat.

The passenger on this 1927 Model H (plowing required a driver and passenger) cranked the blade up and down by hand. He became known as the "wing man" because later plows had a leveling or "wing" blade on the right-hand side that had to be elevated to keep from hitting mailboxes and signs. With the driver controlling the "live" front axle and the passenger operating both blades, it is no wonder they could get by without a cab heater.

This power company's utility truck has tire chains front and rear and a roof-mounted spotlight.

A Model H with a 500 gallon milk tank. Fresh milk had to reach the dairy regardless of road conditions in the 1920s.

Here is another "Oshkosh Express." Note the two-digit telephone number.

This truck is typical of the type purchased by many counties for road-maintenance work. The use of solid tires was common for this type of truck in the 1920s where road speeds were slow.

This early experimental Oshkosh truck was developed with the US Army in the 1920s. Note the wide-track wheels that ran the full width of the truck for maximum flotation.

A Model H cargo-delivery truck owned by Copps, a Wisconsin retail grocer still in business after 100 years.

This delivery truck also was used to clear snow-covered roads. Many Oshkosh owners used their trucks in dual roles.

Construction trucks often had open cabs for maximum visibility. Perhaps the driver is seated on the hood for warmth.

Next page
A Model TR "Earth Mover" tractor. Co-founder B. A. Mosling is in the foreground pointing to the 150hp engine. An air-powered steering device provided four-wheel steer.

These new Model H dump trucks await delivery to a construction contractor.

A Model TR demonstrates its earth-hauling capabilities.

Next page
This 1932 Model F is a good representation of classic 1930s automotive styling.

A TR is loaded with a shovel at a gravel pit.

A photographer frames a new Model F in a natural setting.

F models had vent wings along the side of the hood. They were operated by hand for engine compartment cooling in warm weather. The windshield opened forward for cab ventilation. By this time, all trucks were equipped with cab heaters that made winter operation more comfortable.

A mountain pass in the State of Washington is cleared with a Model G. These trucks, with their longer hoods and larger engines and transmissions, were a larger version of the Model F.

This and all other snowblowers are powered by an engine mounted on the back of the truck. Snowblowers are geared to travel as slow as ½mph, allowing them to chew away high, crusted snow banks.

An Oshkosh with a snowblower shoots snow over a bank. Rotary snowblowers were used to cut through clogged roads and cut back high snow banks. They were also used on airports to keep snow banks from forming along runways.

The polished, cast-aluminum radiators and larger cabs were the most notable differences from early trucks. The Model F was powered by a six-cylinder Hercules gasoline engine.

A Model F truck with a low-boy trailer hauled construction equipment to job sites.

A Model F is used as a concrete-mixer chassis, providing all-wheel-drive delivery over soft ground conditions.

A stylish Model J truck with streamlined front fenders.

Here's another early venture into the all-wheel-drive concrete-carrier market. Built in the late 1940s, it is derived from a snow-removal chassis.

A Model J equipped with a garbage body could double as a snow-removal vehicle.

Next page
Bird's-eye view of a Model F with cab and hood removed.

71

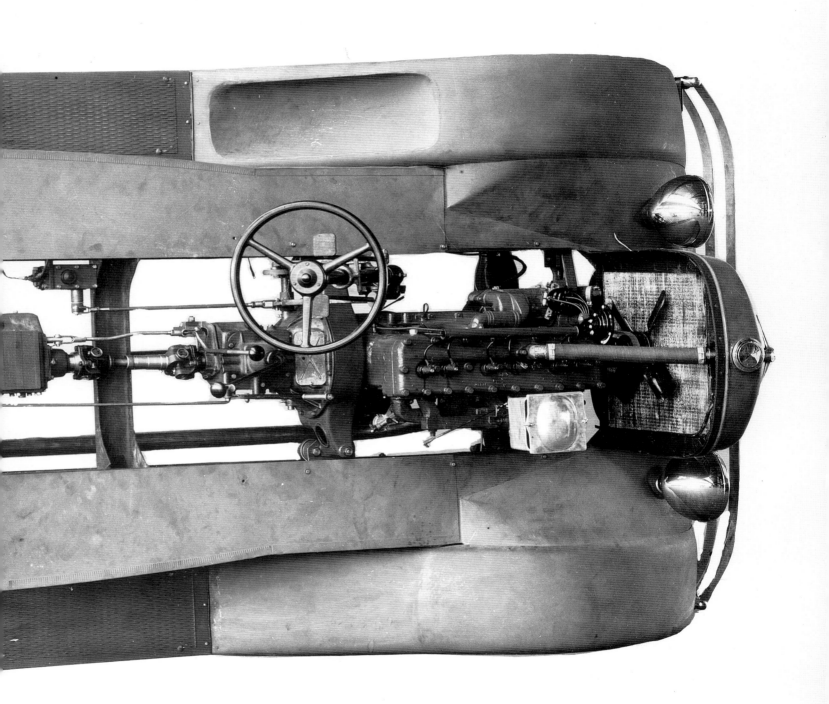

A W-Series snowplow is loaded on a US Air Force C-82 (Flying Boxcar) in February 1949. The truck was destined for a western Indian reservation that was snowbound. The snowplow opened reservation roads for deliveries of food and supplies.

A special W-Series built for the city of Oshkosh in 1940 assured getting to fires during the worst blizzards.

Next page
Another W-Series utility truck, built in the 1950s. All-wheel-drive assured travel into rough terrain for the installation of power lines.

This Model W-212, equipped with a 1,000-gallon tank, was purchased by the US Atomic Energy Commission.

This Model W-1600 vehicle was used by the oil-field industry in many parts of the southwestern United States.

Previous page
Top
A 1950s W-Series 6x6 with a crew cab and utility body.

Center
The Model W-1600 was used to drill for oil. All-wheel-drive assured movement of heavy loads into rough terrain.

Bottom
This W-Series utility truck is equipped with a special five-man cab.

This 1957 fire truck featured a Hall Scott gasoline engine. The giant truck was always the first to arrive at fire scenes in stormy weather.

These Model W-800 series trucks were used by many counties and townships for snow removal and road maintenance.

This W-2205 tractor trailer was used to haul heavy roadbuilding equipment. It also was used for snowplowing. A large concrete weight block was placed on the back of the truck to increase traction since there was no dump body.

This W-Series truck equipped with a rotary snowblower was used at General Mitchell Field in Milwaukee in the 1950s.

Here is a Model W-2200 with side-dump trailer operating in an open pit iron-ore mine. The engine is powered by LP gas.

The US Navy used a fleet of these W-800 series trucks for roadbuilding and snow removal at bases.

A W-Series truck for the US military was used as a bomb-disposal vehicle as well as for wrecker service.

Only one W-3000 was built and this is it. The tractor-trailer was constructed in 1967 for use in mining.

This W-800 series was one of the company's first Aircraft Rescue and Fire Fighting (ARFF) trucks. The turret dispensed fire-extinguishing foam.

This experimental tractor/trailer unit increased carrying capacity. The mixer could be moved forward on the trailer for better weight distribution when traveling, then moved back for concrete discharge.

This W-Series with a garbage body also was used for snow removal in Yonkers, New York.

Here, a Model 1834 hauls concrete block.

All-wheel-drive was an important feature on this W-Series rural fire truck, built in the 1960s. Getting through snow in winter and mud in spring was vital to many rural departments.

This 6x6 concrete carrier with extra-large front tires delivered ready mix where a conventional rear-wheel-drive truck would have gotten stuck.

A US Air Force MB-2 tow tractor tows a vintage B-29 bomber at an Experimental Aviation Association convention and show in Oshkosh.

A US Air Force WT-2206 with roll-over plow. The WT-2206s were equipped with Hall Scott engines and automatic transmissions.

Seventy-five of these W-1600 crane trucks were made for the US Navy in the late 1950s for aircraft rescue.

A US Air Force U-30 four-wheel steer tow tractor.

Over 1,000 of these large, high-speed WT-2206 trucks were delivered to the US Air Force beginning in the 1950s. Their mission was to keep runways clear of snow for Strategic Air Command bombers.

Oshkosh sold many of its A-Series chassis to fire equipment manufacturers in the 1960s and 1970s.

Here's an L-Series chassis with special railroad-track maintenance body.

A cab-over 6x6 G-Series was designed in the 1960s for oil-well servicing worldwide.

When the reverse-slope windshield was introduced, the C-Series became the M-Series. Here is one being used as a block truck.

The 6x4 E-Series truck was developed for the export market in the 1960s. Here a truck hauls cane in its body along with the trailer it pulls.

The 6x6 C-Series refuse hauler could get in and out of soft landfills when dumping garbage.

The low-profile cab of the L-Series was designed to be used for older fire houses with low entrances.

Many Oshkosh trucks are used to haul concrete blocks onto construction projects.

In 1976, this concrete carrier's paint scheme paid tribute to the US Bicentennial. The tag axle was used to increase loads.

A 6x6 C-Series concrete mixer discharges concrete in the Carolinas.

This 6x6 was made in 1960 for an independent logger in Columbia Falls, Idaho, and still operates today. He came to the factory and watched its construction.

This M-Series chassis features a self-loading log-cradle body. All-wheel-drive assured mobility into and out of the forest.

This one-of-a-kind 8x8 was built in the early 1960s. It was powered by a 500hp engine for transporting oversize loads in California.

The D-Series with its driving, steerable front tandem maximized the number of yards of concrete the truck could haul and increased off-road mobility.

Hilltop Concrete in Cincinnati operated several hundred Oshkosh concrete carriers.

Many R-Series 6x4 tractors were used to haul construction equipment. Most R-Series trucks were used overseas where their extra-heavy-duty design endured the rugged conditions found in many areas of the world.

Previous page
The Model 40K aircraft loader had a 40,000lb capacity. Its low profile permitted the vehicle to roll beneath the wings of parked aircraft.

A number of these K-Series tractors were built in the 1980s for Goodyear Aerospace to pull trailers with missile stages from the factory to the assembly line. The 6x6 was powered by a 600hp engine. Fully loaded, it weighed up to 125 tons and could travel at high speeds on its fifty tires.

Here's an R-Series 6x4 tractor hauling double-bottom trailers in Australia. The large front bumper protects the tractor from slow kangaroos.

This R-Series 6x4 hauls sugar cane in Peru. The inside of the trailer was lined with a chain net. The sugar cane was unloaded at the refinery by a crane lifting the netted bundle.

These Model U-44-L units were sold to the US Navy to drill post holes. They had solid suspension for stability while drilling. The cantilevered engine design provided better weight distribution.

A dozen of these U-44-L trucks were made for the Wisconsin Telephone Company in the late 1960s. The units were used to drill holes and place poles. The passenger sat behind the driver.

The B-Series was designed with the engine in the back so the mixer chute could extend over the low-profile cab. Time was saved and better service provided because the operator could drive the truck and operate the chute at the same time as concrete was placed.

An Oshkosh tractor pulls a load of logs in Southeast Asia.

This S-Series truck is equipped with an Oshkosh brand mixer and high-lift tag axle. The S-Series was an updated version of the B-Series and featured a fiberglass engine cover.

An R-Series 6x4 and its Caterpillar D-9 tractor await loading on the way to Saudi Arabia in the late 1960s.

This F-Series is equipped with pusher and tag axles for maximum legal payload. All-wheel-drive also is important to assure delivery of concrete blocks exactly where the customer wants them.

Next page
The F-Series 6x6 concrete carrier is an especially popular model in the Chicago area. A variety of wheelbases and configurations are necessary on concrete carriers to comply with different state weight laws.

Four F-Series trucks are used to move this massive desalination plant off a ship in Saudi Arabia.

This high speed PA-Series clears a runway at a northern Michigan airport.

Left
This 6x6 truck is equipped with a runway deicer to keep airport pavement ice free.

Previous page
Top
A fleet of H-Series rotary snowblowers arrives at O'Hare Field in Chicago in the late 1980s. Keeping airports open during snow storms is the critical job of these vehicles.

Center
This HB-Series vehicle is equipped with a runway sweeper. The cab provides optimum visibility.

Bottom
A Model HB snowblower can blow 3,000 tons an hour. The four-wheel-steer vehicles are sold to the US Air Force and to commercial airports.

Model J trucks are linked to haul a generator turbine to a dam site in Brazil.

A P-Series with a rollover plow and double wings clears an airport runway.

A P-Series pushes back a snow bank in central Wisconsin. The tag axle is lifted for a short-turning radius and maximum weight on the driving axle while plowing. In the summer, the axle is lowered for maximum carrying capacity hauling gravel.

The multi-purpose HB can be equipped with either a snowblower head, plow blade or broom.

A Model J receives a final check by a technician before delivery to a heavy-haul customer.

This J-Series concrete carrier has special tires for construction work in the desert.

Top
This J-Series unit is ready for overseas shipment to a desert for hauling equipment to oil-field drilling sites.

Center
This US Air Force P-15 ARFF vehicle is being loaded into a C5A transport. The 8x8 had engines mounted front and rear to power each of the tandems.

This Model J oil-field servicing truck has wide, high-flotation tires to travel over the powderlike sand found in the interior of China.

A commercial version of the military P-19 ARFF vehicle. It has a 1,000 gallon water capacity and, like most ARFF vehicles today, center steering.

This is a commercial version of a US Air Force Model P-4, 1,500 gallon capacity Aircraft Rescue and Fire Fighting (ARFF) vehicle built in the early 1970s. ARFF cabs and bodies are made of aluminum for light weight to speed acceleration.

A commercial version of the P-15 which carried 6,000 gallons of water. Fire fighting foam was discharged from the roof turret.

The high-mobility Model DA-Series ARFF unit was derived from a US Marine Corps tactical truck. The articulating joint at the center of the vehicle maximized off-runway mobility because it allowed the unit to "duck walk" through soft ground.

This is a commercial version of the US Navy MB-1, built in the early 1970s.

Next page
This 4x4 T-Series ARFF vehicle carried 1,500 gallons of water. High-flotation tires assure high off-runway mobility.

The Marines called this truck the LVS for Logistics Vehicle System. A flatbed mounted with a shipping container was one of many rear-section types.

The LVS can operate in 5ft of water. This one is configured as a cargo truck with a rear-mounted crane.

*Previous page
This 6x6 T-Series ARFF vehicle has a 3,000 gallon capacity. Units are equipped with both roof and bumper turrets.*

Equipped with sand tires for desert operation, this LVS shows the superior turning radius provided by the center articulation joint. The front axle also steers. This 8x8, which was built for the US Marine Corps, could uncouple at its articulation joint. The rear section is interchangeable with other types of rear sections. Uncoupling also made helicopter transport possible.

Previous page
Top
Oshkosh has built over 13,000 of these 8x8 Heavy Expanded Mobility Tactical Trucks (HEMTTs) for the US Army. This 2,500 gallon tanker is one of several versions that played a key role in the success of Operation Desert Storm, keeping vehicles and helicopters fueled.

Center
This HEMTT wrecker can recover just about any type of wheeled military vehicle in the US Army's inventory.

Bottom
A HEMTT cargo truck and trailer unloads rocket pods.

The US Army uses the 500hp Palletized Load System (PLS) truck for quick loading and unloading of military cargo. One person can load or unload the flatrack from inside the cab. An arm with a hook hydraulically lifts the load on and off.

The US Air Force uses the R-11 aircraft refueler to transport aircraft fuel. The 6,000 gallon capacity 6x4 units also can defuel aircraft.

Previous page
The 6x6 HET is most often used to transport tanks.

The quick-loading PLS is a new generation of heavy Army truck that permits transport of material much faster and with fewer people. It is a 10x10 vehicle with a central tire-inflation system. The flatrack is demounted so the truck can return for another load. In addition to tandem front steering, the last axle steers for maximum maneuverability. It carries 16½ tons and pulls a trailer with its own 16½ ton payload.

The US Army M-911 Heavy Equipment Transporter (HET) was first built by Oshkosh in the 1970s and saw extensive use during Operation Desert Storm.

The Oshkosh refuse compactor/recycler is two trucks in one. It can do the job with one operator that previously required two single-function trucks. The vehicle can be operated from either side of the low-profile cab.

The Army's newest Oshkosh Heavy Equipment Transporter is the M-1070. Like the PLS, it has 500hp and a central tire-inflation system. It has a five-man cab and a rear steering axle to minimize the turning circle. The prime mission of the vehicle is the transport of the M1A1 US Army tank.

Afterword

Looking to the Future

Bernhard A. Mosling and William A. Besserdich, the businessman and the engineer, would be amazed at the size and diversity of the company they began seventy-five years ago. Yet the truck-manufacturing pioneers would recognize the same dedication to technical advances, pride in craftsmanship, cost consciousness, and commitment to customers that helped Oshkosh Truck Corporation survive and thrive.

From humble beginnings, Oshkosh Truck Corporation has grown by applying ingenuity and the latest technology to the needs of its customers. On pavement or off, Oshkosh Truck Corporation will continue to meet unique transportation needs through specialized design. Their products will, in the future as in the past, "Go Anywhere the Wheels Touch the Ground."

Index

A-Series, 89
A. O. Smith Company, 13
Allison automatic transmissions, 25, 34

B-Series, 23, 103-104
Besserdich, William R., 9, 14
Brown-Lipe Model 35 transmissions, 12

C-Series, 22-23, 92-94
Caterpillar D-9, 103
Cummins diesel engines, 20

D-Series, 23, 96
Detroit Diesel V-8s, 29, 34

E-Series, 25-26, 92

F-Series, 22-24, 105

G-Series, 91
General Mitchell Field, 79
Goodyear Aerospace, 100
Goodyear tires, 13

H-Series, 109
Hall-Scott gasoline engines, 20, 25, 78
HB-Series, 109, 111
Hilltop Concrete, 97

J-Series, 17, 24, 112-113

K-Series, 100

L-Series, 90, 93
Le Roi engines, 12

M-Series, 92, 95
Mackie, R. W., 15

Models:
 45-55, 20
 50-50, 20-22
 1832, 22
 1834, 84
 2200, 21
 A, 13-14, 45, 53
 B, 14, 49-50, 55
 C5A, 31
 DA, 115
 F, 14, 64, 66-67, 69-71
 FB, 15-16
 FC, 15
 G, 67
 H, 14-15, 53, 57-58, 60, 64
 J, 70-71, 110
 M-911, 31, 33, 37, 125
 M-1070, 37, 126
 MB-1, 29, 115
 MB-2, 31, 33
 MB-5, 29
 P-4, 30, 114
 P-15, 29-30, 115
 P-19, 29, 31, 114
 R-11, 34, 36, 124
 TR, 17, 61, 64, 66
 U-30, 31, 87
 U-44-L, 102
 W-212, 77
 W-700, 19
 W-800, 78, 81, 83
 W-1600, 19-20, 77, 86
 W-2200, 20, 81
 W-2205, 79
 W-3000, 82
 WT-2206, 25-27, 86-87
Mosling, Bernhard A., 9, 12, 14-15, 20, 45, 61
Mosling, J. P., 12

O'Hare Field, 109
Old Betsy, 9, 12
Operation Desert Storm, 121
Oshkosh Motor Truck Company, 15
Oshkosh Motor Truck Manufacturing, 12, 14

P-Series, 110-111
PA-Series, 109

R-Series, 24-25, 27, 100-101, 103

S-Series, 24, 103
Schneider, J. H., 13
Schwarzkopf, H. Norman, 34
Standard Oil Company, 50
Stone, Michael P. W., 34

T-Series, 115, 119
Tuttle, Gen. William G. T., Jr., 34

US Air Force, 29-31, 33-34, 36-37, 74, 86-87, 109, 113
US Army Heavy Expanded Mobility Tactical Truck (HEMTT), 35
US Army Material Command, 34
US Army, 31, 33, 37-38, 125
US Atomic Energy Commission, 77
US Marine Corps, 31, 115, 119, 121
US Marines Logistics Vehicle System (LVS), 36
US Navy, 29, 81, 102, 115

W-Series, 19-29, 74, 77, 79, 81, 83-85
Waupaca County Highway Department, 53
Wisconsin Duplex Auto Company, 9, 45
Wisconsin Hemp Company, 53
Wisconsin Telephone Company, 102